ANIMALS OF THE FOREST

Woodpeckers

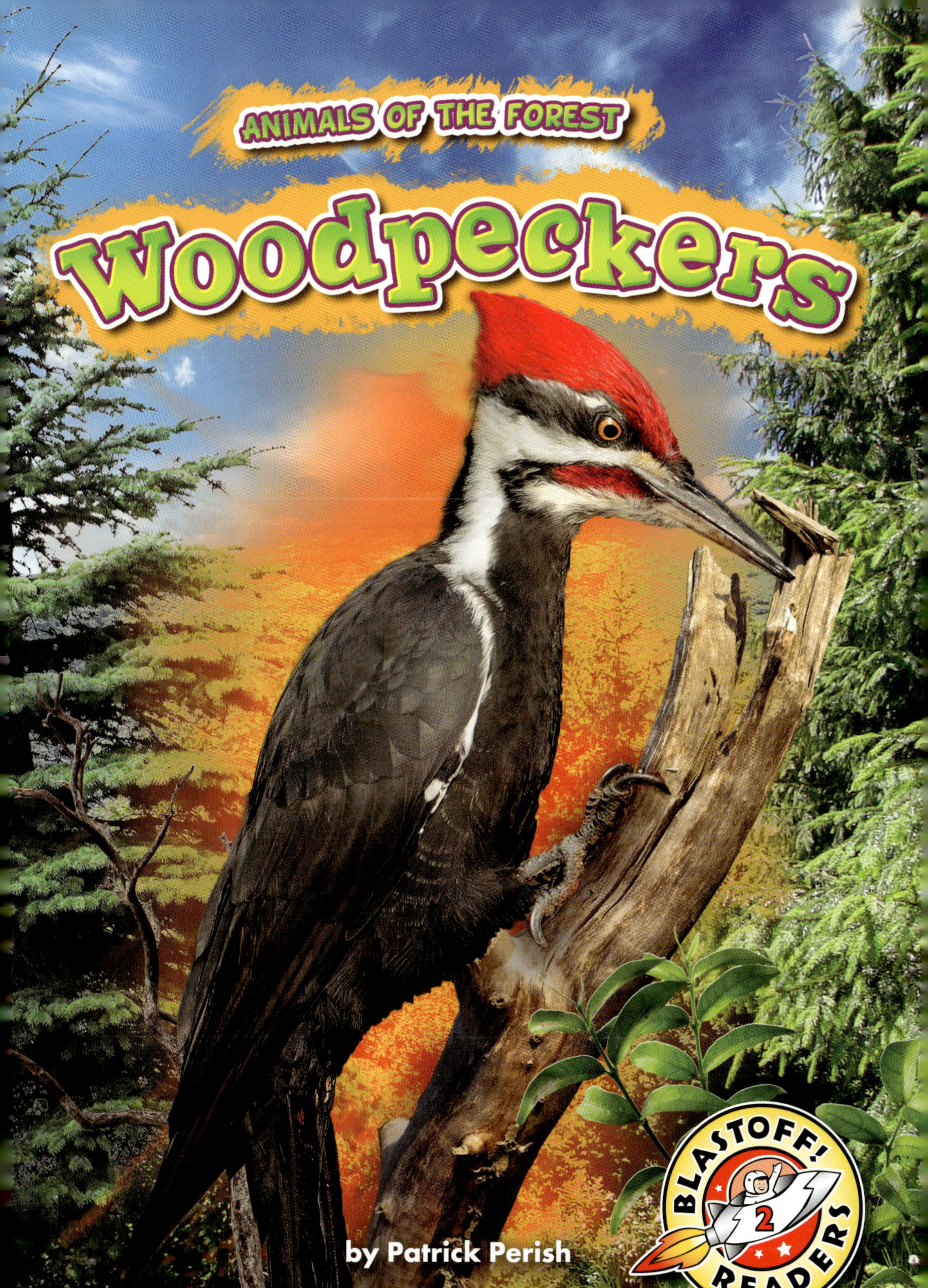

by Patrick Perish

BLASTOFF! READERS 2

BELLWETHER MEDIA • MINNEAPOLIS, MN

Blastoff! Readers are carefully developed by literacy experts to build reading stamina and move students toward fluency by combining standards-based content with developmentally appropriate text.

Level 1 provides the most support through repetition of high-frequency words, light text, predictable sentence patterns, and strong visual support.

Level 2 offers early readers a bit more challenge through varied sentences, increased text load, and text-supportive special features.

Level 3 advances early-fluent readers toward fluency through increased text load, less reliance on photos, advancing concepts, longer sentences, and more complex special features.

★ **Blastoff! Universe**

This edition first published in 2022 by Bellwether Media, Inc.

Library of Congress Cataloging-in-Publication Data

Names: Perish, Patrick, author.
Title: Woodpeckers / by Patrick Perish.
Description: Minneapolis, MN : Bellwether Media, 2022. | Series: Blastoff! readers: Animals of the forest | Includes bibliographical references and index. | Audience: Ages 5-8 | Audience: Grades 2-3 | Summary: "Relevant images match informative text in this introduction to woodpeckers. Intended for students in kindergarten through third grade"– Provided by publisher.
Identifiers: LCCN 2021004013 (print) | LCCN 2021004014 (ebook) | ISBN 9781644875094 (library binding) | ISBN 9781648344176 (ebook)
Subjects: LCSH: Woodpeckers–Juvenile literature.
Classification: LCC QL696.P56 P47 2022 (print) | LCC QL696.P56 (ebook) | DDC 598.7/2-dc23
LC record available at https://lccn.loc.gov/2021004013
LC ebook record available at https://lccn.loc.gov/2021004014

Editor: Kieran Downs   Designer: Josh Brink

Printed in the United States of America, North Mankato, MN.

Table of Contents

Life in the Forest

downy woodpecker

Woodpeckers are made
for life in the trees.

These birds can be found around the world. They live well in the forest **biome**.

Downy Woodpecker Range

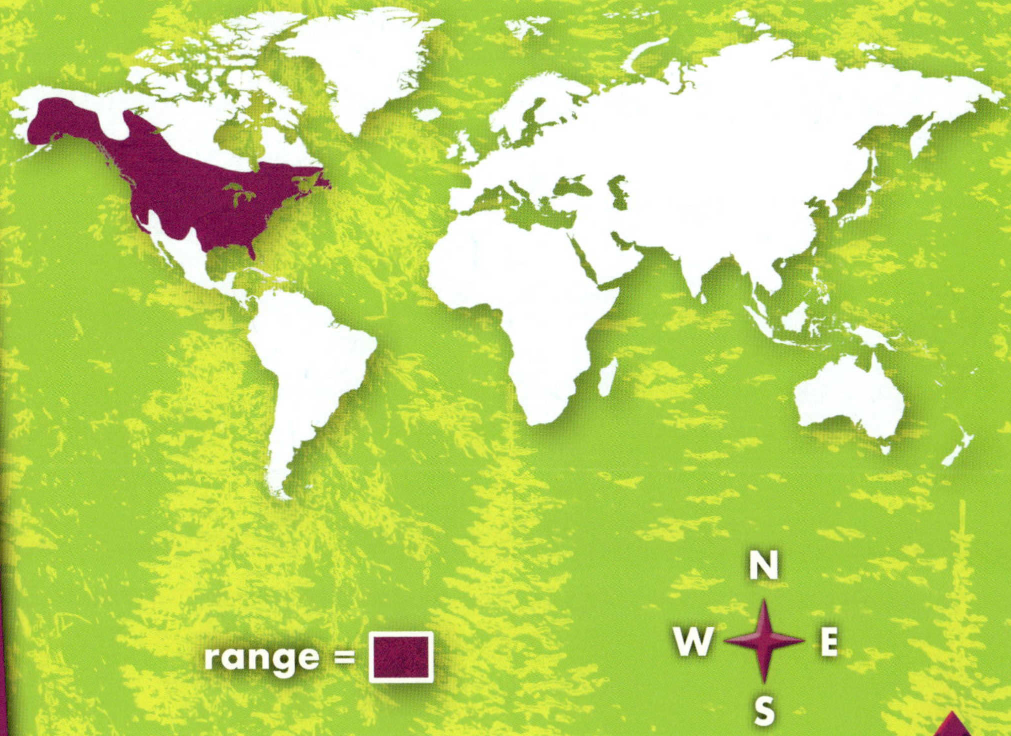

Woodpeckers have beaks that are **adapted** for pecking trees. Their sharp beaks chip away wood.

Small feathers keep woodchips out of the birds' **nostrils**.

Woodpecker feet have two forward toes and two backward toes. The toes grip tree trunks while the birds are pecking.

red-bellied woodpecker

Strong tail feathers help them balance.

Most woodpeckers have long tongues. Some tongues have sticky **saliva**.

Others have small **barbs**. These help woodpeckers pull food from holes.

Special Adaptations

Busy Beaks

Woodpeckers chip holes in trees with their beaks. They use the holes for nests.

Their eggs stay safe
high above the forest floor.

Woodpeckers also peck to **communicate**. They rap on trees, posts, or even houses!

These sounds help them find **mates** or mark **territory**.

Downy Woodpecker Stats

conservation status: least concern

life span: up to 5 years

Forest Foods

middle spotted woodpecker

Woodpeckers eat a lot of **insects**. They stick their tongues inside the holes they create to reach food.

They also catch insects on tree trunks or branches.

Woodpecker Diet

carpenter ants

white pine pinecones

poison ivy berries

Many woodpeckers also eat tree **sap**. They peck small holes into trees to lick it up.

Others prefer seeds or fruit. Some woodpeckers store food to eat later.

Woodpeckers keep insect pests under control. Squirrels, owls, and other animals use old woodpecker nests.

These birds help keep forests healthy!

21

Glossary

adapted—well suited due to changes over a long period of time

barbs—small hooks

biome—a large area with certain plants, animals, and weather

communicate—to share information and feelings

insects—small animals with six legs and hard outer bodies; an insect's body is divided into three parts.

mates—partners

nostrils—the two openings of the nose

saliva—a watery fluid in the mouth that breaks down food

sap—juices flowing inside trees

territory—the land area where an animal lives

To Learn More

AT THE LIBRARY

Murray, Julie. *Woodpeckers.* Minneapolis, Minn.: Abdo Publishing, 2020.

Sabelko, Rebecca. *Pileated Woodpeckers.* Minneapolis, Minn.: Bellwether Media, 2019.

Statts, Leo. *Woodpeckers.* Minneapolis, Minn.: Abdo Zoom, 2018.

ON THE WEB

FACTSURFER

Factsurfer.com gives you a safe, fun way to find more information.

1. Go to www.factsurfer.com.
2. Enter "woodpeckers" into the search box and click 🔍.
3. Select your book cover to see a list of related content.

Index

The images in this book are reproduced through the courtesy of: Tim Zurowski, cover (hero); Michiru13, cover (background); Alex Stemmer, CIP; Brian A Wolf, p. 4; Harry Collins Photography, p. 6; Mircea Costina, pp. 7, 12, 20; Bildagentur Zoonar GmbH, p. 8; Carla Mason, p. 9; Ken Griffiths, p. 10; 6381380, p. 11; FotoRequest, pp. 11 (four toes), 13; David McManus, p. 14; HeavyT Photography, pp. 14-15; serkan mutan, p. 16; Henrik Larsson, p. 17 (carpenter ant); Konmac, p. 17 (white pine pincecone); Rob and Heather Hercules, p. 17 (poison ivy berries); Wild Birds, Alius Imago/ Alamy, p. 18; Hayley Crews, p. 19; David Savile, pp. 20-21; rck_953, p. 23.